Francis E. Nipher

An Introduction to Graphical Algebra

For the Use of High Schools

Francis E. Nipher

An Introduction to Graphical Algebra
For the Use of High Schools

ISBN/EAN: 9783337158576

Printed in Europe, USA, Canada, Australia, Japan

Cover: Foto ©Paul-Georg Meister /pixelio.de

More available books at **www.hansebooks.com**

AN INTRODUCTION TO
GRAPHICAL ALGEBRA

FOR THE USE OF HIGH SCHOOLS

BY

FRANCIS E. NIPHER, A.M.

*Professor of Physics and in charge of Electrical Engineering
in Washington University*

NEW YORK
HENRY HOLT AND COMPANY
1898

AN INTRODUCTION TO
GRAPHICAL ALGEBRA

FOR THE USE OF HIGH SCHOOLS

BY

FRANCIS E. NIPHER, A.M.

Professor of Physics and in charge of Electrical Engineering in Washington University

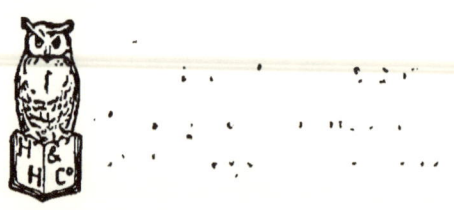

NEW YORK
HENRY HOLT AND COMPANY
1898

Copyright, 1898,
BY
HENRY HOLT & CO.

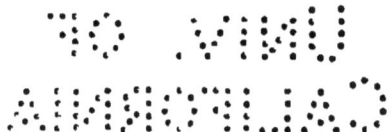

ROBERT DRUMMOND, PRINTER, NEW YORK.

PREFACE.

The author has been asked to present a paper for the consideration of the Trans-Mississippi Educational Convention, to be held at Omaha on June 28, 29, and 30, 1898. The subject assigned was the greater efficiency of science instruction.

It did not seem desirable to merely suggest in a general way what appears to the author the most desirable change in high-school instruction in order to increase the efficiency of college work. The aim has been rather to show how such changes may be accomplished without radical departures from present methods.

By injecting here and there into the ordinary instruction in algebra such material as is found in the following text, new meaning will be given to the operations involved in the solution of equations, and new interest in the subject may be aroused. The study of geometry by Euclid's method requires a large amount of time which for the average student might be more usefully employed. The study of algebra and geometry as wholly distinct subjects having no relation to each other, gives to the pupil a false idea of the intellectual situation of to-day. The scientific investigator has, since Newton's time, very largely abandoned Euclid's methods as applied to scientific investigation. It does not therefore follow that Euclid's geometry should be banished from our schools, but it does seem proper to consider whether some of the time given to it might not be more usefully spent in elementary analytical geometry or graphical algebra.

The devotees of geometry and mathematics often seem impatient that these subjects should be studied except for themselves alone, and for the intellectual enjoyment and mental development which they afford. But while this may be a very proper mental attitude for such men, it does not follow that we should all adopt this view of the matter. The laws of the physical universe are equally worthy of human consideration. And these laws are most impressively presented to the mind in the symbolic language of algebra or the graphical language of geometry.

We may forgive a civil engineer if he confines his admiration to the beauties of a properly designed truss; nevertheless we may find new beauties in a graceful building of which the truss forms a useful and necessary part.

The cross-ruled paper needed for the student's use in graphical solutions may be obtained from any dealer in draughtsmen's supplies.

F. E. N.

St. Louis, Mo., June, 1898.

AN INTRODUCTION
TO
GRAPHICAL ALGEBRA.

WHEN two physical quantities are so related to each other that a definite change impressed upon one is always accompanied by a corresponding definite change in the other, such relation is called a physical law.

The statement that the weight of a cylindrical rod of iron is proportional to its length is a simple example of a physical law.

Every such law may also be stated in an algebraic equation.

It may also be completely expressed in the graphical language of geometry.

The examples that follow are designed to show the physical and geometrical significance which may be attached to the ordinary equations usually discussed in the algebra of the preparatory schools.

1. *The weight of a wire or rod is directly proportional to its length.*

Let a represent the weight per foot of the wire or rod. Then the weight m of l feet will be

$$m = al. \qquad (1)$$

This equation asserts that m and al are equal. If the wire is of uniform size, then it must follow that if one piece of it is twice as long as another it must have twice the weight, or that m increases in a direct ratio with l. However the values of l may differ in different pieces of the same kind of wire, the weights also so differ that $m \div l$ is always the same and is equal to a. The equation may be written

$$\frac{m}{l} = \frac{a}{1},$$

or $\qquad m : l = a : 1.$

The direct ratio represented in (1) between m and l is the most common of all physical laws.

The value of merchandise is directly proportional to the amount. The compensation due to a workman is proportional to the time-interval of his service. The distance traversed by a body moving at uniform speed is proportional to the time. Within proper limits, the yielding of a bridge is proportional to its load, etc.

2. If squares of metal be cut from a uniform sheet, *the weight is proportional to the square of the length of the side.*

Let b represent the weight per square foot of the metal sheet, and let $l =$ the length of one side of any square. Then the weight m of any square is

$$m = bl^2. \qquad (2)$$

This equation teaches that if l^2 be doubled (that is to say, if the area of the plate be doubled), the weight will be doubled. If, however, l be doubled, the weight will become four times as great.

Let us assume by way of example that $a = 8$ and $b = 2$. The wire is in that case to be a rod weighing 8 lbs. per foot, and the metal sheet is to weigh 2 lbs. per square foot. Equations (1) and (2) then become

$$m = 8l, \qquad (1)'$$

$$m = 2l^2. \qquad (2)'$$

We may now compute the weights m of rods of various lengths l, and of squares of such metal having various lengths l. They are given in Table 1.

Let us now assume that l has the same numerical value in the two equations. Physically this would mean that we are to associate with any rod l feet in

TABLE I.

l.	$m =$ Weight of	
	Rods.	Squares.
0	0	0
1	8	2
2	16	8
3	24	18
4	32	32
5	40	50
6	48	72
7	56	98
8	64	128
9	72	162
10	80	200

length a square having a side of l feet. Then in general m would not be the same in the two equations. If, for example, $l = 10$, the weight of the rod would be 80 lbs., while the weight of the square would be 200 lbs., as the table shows. A similar assumption of equality in the values m would enable us to eliminate m, but we must distinguish between the two values l thus:

$$2l'^2 = 8l'.$$

If, however, we also simultaneously assume that the l of one equation is the same as the l of the other, we may write the last equation

$$2l(l-4) = 0.$$

Here is an equation in which the product of two factors is zero. This will be possible if either factor is zero. That is to say, the m of one equation will be the same as the m of the other equation *and* the l of one will be the same as the l of the other, either when $2l = 0$, and therefore when $l = 0$, or when $l - 4 = 0$, or $l = 4$.

In other words, $2l^2$ will equal $8l$ either when $l = 0$ or when $l = 4$. When $l = 0$, the m of each equation is zero, and when $l = 4$, it is 32 lbs., as the table of values shows.

In order to represent geometrically the nature of these two equations, involving these two laws, let us lay off along a horizontal direction the lengths l ranging from 0 to 10 (Fig. 1), and forming an axis or scale of lengths. At right angles to this axis lay off any other scale for the values m. In the figure this scale ranges from 0 to 100. The scale value of one inch measured along the axes may be the same for m and l, or the scales may be different, as convenience may determine.

Corresponding to various values l, along the horizontal scale, lay off distances m as given in the table. If the first column m be taken, the points so determined will lie in a straight line marked (1)' in the figure. The vertical distance between the horizontal axis l and the inclined line (1)' is proportional to its distance l from the intersection at the origin. These vertical lines may be considered the bases of triangles having a common vertex at the intersection. The bases of similar triangles are proportional to their altitudes.

If the second column of the table be similarly diagrammed with the corresponding values of l, a

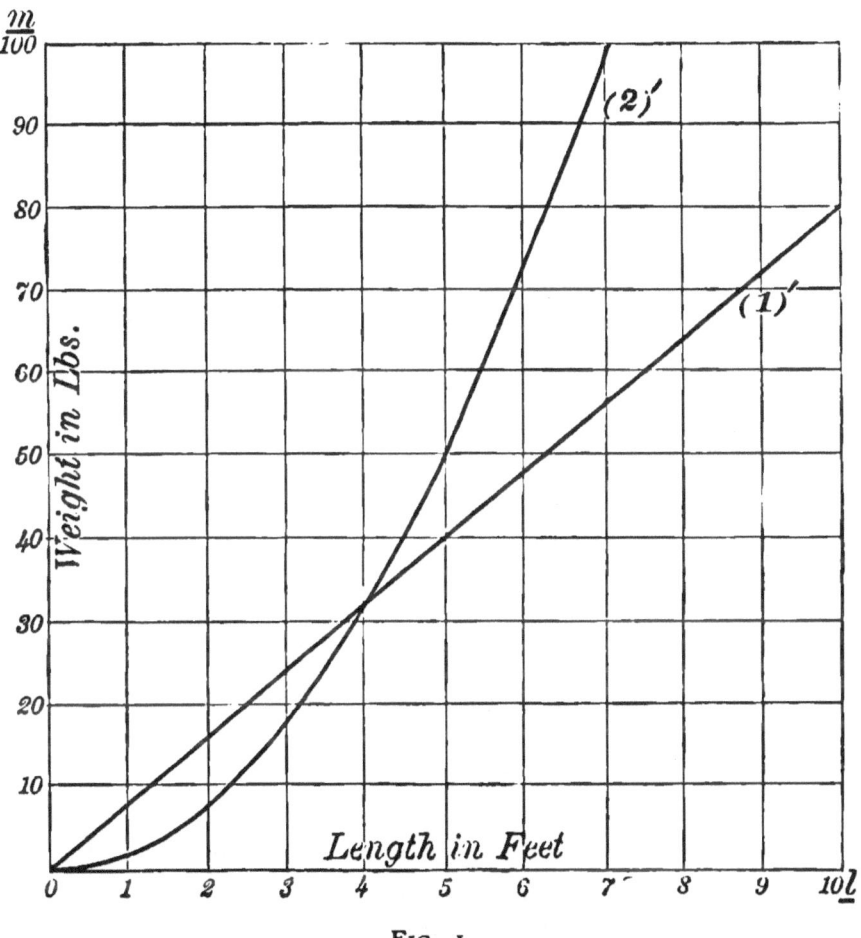

FIG. 1.

curve marked (2)' is obtained as shown in the figure. In this case the vertical distance from the horizontal axis to the curve is proportional to the square of l, instead of to the first power as in the other case.

Equation (1), where a is any constant, is said to be the equation of a straight line. If $a = 8$, as in (1)', the equation is then the equation of a particular straight line, whose slope $\dfrac{m}{l}$ is 8, and which is shown in Fig. 1. Similarly, equation (2)' is the equation of the curve shown in that figure. The points on the straight line are located with reference to the two axes on which the scales are marked, exactly as points on a map are located by their latitude north or south of the equator, and their longitude east or west of any assumed meridian. If one were to consider a small region a few miles in extent around the point where the meridian of zero longitude crosses the equator, we might locate points by measuring the distances, in feet, east or west of the meridian and the distances north or south of the equator. We could by means of a surveyor's chain and compass find the point whose longitude is east 50 ft. and whose latitude is north 400 ft.

The points whose latitudes are 8 times the longitudes would lie in a straight line. If we call y the latitude and x the longitude, and write the equation

$$y = 8x,$$

we are then confining our attention to a definite series

of points lying in a straight line, if we neglect the curvature and irregularities of the earth's surface. If

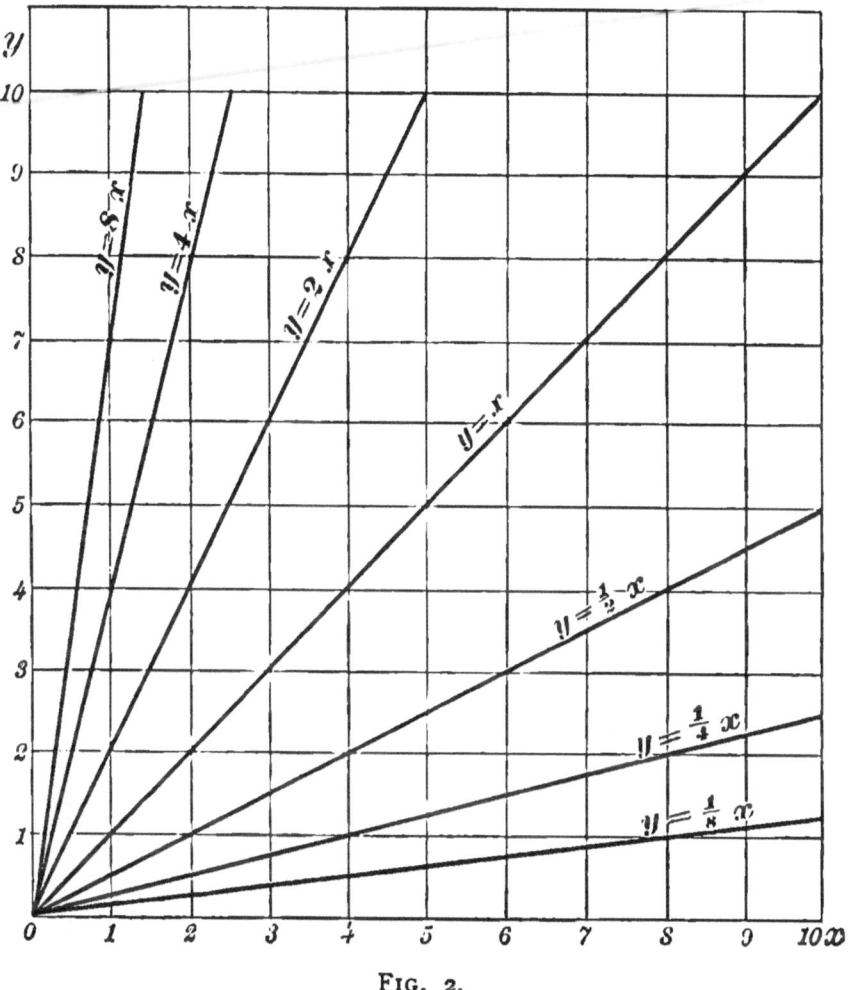

Fig. 2.

we write $y = 4x$, the line of points would be a different one. The slope of the line would be only half as great. The smaller the constant factor, the less steep will be the line. Such lines are shown in Fig. 2.

It is evident that we may in a similar way represent by a line the relation between any other two quantities when one quantity is a certain number of times the other. This is what has been done in Fig. 1. In that case the weight in pounds of a rod is 8 times its length in feet.

If the rod had only half the section, then m would be 4 times l, and the line representing the relation between weight and length would be only half as steep. In a similar way, if the sheet of metal were only one fourth as thick as that represented by (2)', that equation would become $m = \frac{1}{4}l^2$. The curve representing the relation between m and l would be like that shown in Fig. 1, but would lie between that curve and the axis l. This curve may be easily laid down upon squared paper, after a table of values of m for various values l has been computed.

It is evident from what has preceded that equations (1)' and (2)', when separately considered, represent entirely different relations between a weight and a length, and that we may assign any value to l, in either equation, and find m by computation if we know the numerical values of a and b.

The physical statement corresponding to this is that we may compute the number of pounds in any wire, or in any square sheet of metal, if we know in the one case the weight per foot, or in the other case the weight per square foot.

The diagram Fig. 1 also permits us by inspection to determine these values of m for various values l within its limits.

3. Let us suppose that a merchant wishes to issue x copies of a one-page circular letter. He finds that a typewriter will charge him 5 cents a copy. The cost y of x copies will then be

$$y = 5x. \quad \ldots \quad \ldots \quad (3)$$

A printer offers to furnish them at the following rates: \$2.10 for the first hundred and 10 cents extra for each additional hundred.

For the work of printing alone and furnishing paper the printer charges, therefore, $\frac{1}{10}$ cent a copy. In the first 100 copies, 10 cents covers the cost of printing and paper, and therefore the printer must have charged \$2 or 200 cents for setting the type and preparing to print. The cost of x printed copies will therefore be, in cents,

$$y = 200 + \tfrac{1}{10}x. \quad \ldots \quad \ldots \quad (4)$$

We may now compute in a table the cost y, in cents, of any number x of copies.

TABLE 2.

x.	$y =$ Cost of x copies.	
	Typewriter.	Printer.
0	0	200
10	50	201
20	100	202
30	150	203
40	200	204
50	250	205
100	500	210
1000	5000	300

Let us now find the values of x and y in (3) and (4) by elimination. Eliminating y, we have

$$5x = 200 + 0.1x,$$
$$4.9x = 200,$$
$$x = 40.8 +.$$

Hence $\qquad y = 204 +.$

These values of x and y were obtained by assuming that the x and the y of one equation are the same as the x and the y of the other. This value of x is the particular number of copies that would cost the same by one method of production as by the other. The value of y is the cost of this number. For a less number of copies the typewriter method would be the cheaper. When x is zero, the equations assume that the preparation for printing has been made, but none are printed. The cost of preparation with the typewriter method is zero, and by the other method the cost of setting the type is 200 cents.

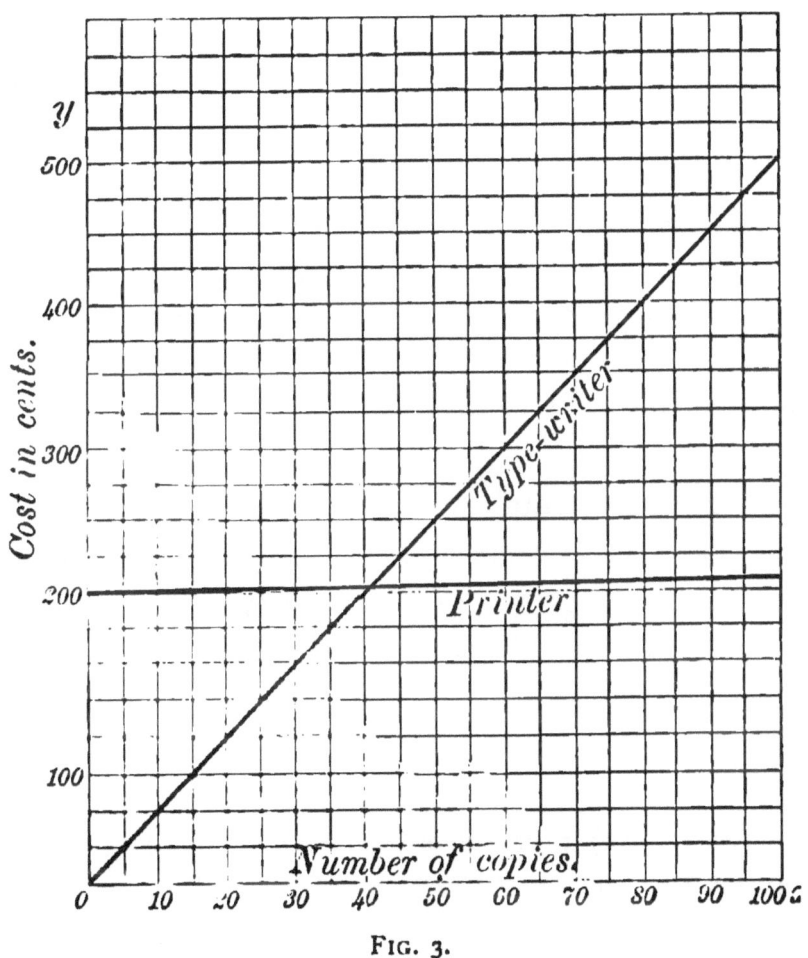

FIG. 3.

These two equations (3) and (4) are also the equations of lines. They are shown in Fig. 3.

The values of x and y just found by elimination

determine the point where these lines intersect. These lines show to the eye that for a small number of copies the typewriter method will be the cheaper, while for a greater number than 40 the printing method will be cheaper. As is shown in the table, the cost of 1000 copies will be $50 by one method and only $3 by the other.

4. The keeper of a notion stand finds that on Jan. 1, 1898, his cash and stock exceed his debts by $20. He is, however, losing each day $2. At the end of x days after Jan. 1st he will be worth

$$y = 20 - 2x. \quad \ldots \quad (5)$$

On the same date another merchant finds that his debts exceed the value of his cash and stock by $15, but he is making a net profit each day of $3. At the end of x days he will be worth

$$y = -15 + 3x. \quad \ldots \quad (6)$$

Equation (5) represents the business career of the first dealer. He will soon have lost all his money. The value of y in that equation will then be zero. We can find when this will happen by making $y = 0$ in that equation. It then becomes

$$0 = 20 - 2x.$$

This shows that $x = 10$ when $y = 0$. His wealth will become zero on Jan. 11th, or ten days after Jan. 1st. If he continue in business he will be going into debt.

If we make $y = 0$ in equation (6), we have

$$0 = -15 + 3x,$$

or $\quad x = 5.$

This shows that on Jan. 6th, the second dealer will be out of debt.

If we find y and x in equations (5) and (6), we have

$$20 - 2x = -15 + 3x,$$
$$5x = 35,$$
$$x = 7.$$

This shows that seven days after Jan 1st these men will have equal fortunes. Making $x = 7$ in either (5) or (6), we find that $y = +6$, which will be what each will be worth at that time.

By assigning consecutive values to x in (5) and (6) we find the fortune of these men on successive days after Jan. 1st as shown in Table 3.

The sinking and rising fortunes of these men are also shown in Fig. 4. The line marked (5) in Fig. 4 starts at a distance 20 above the x-axis, and sinks to that axis where $x = 10$. Prior to that time, the distance from the x-axis up to the inclined line represents his net resources or his wealth. After that time, the distance from the x-axis down to the inclined line represents his net liabilities. His wealth is then a negative quantity. In a similar way the line marked (6) shows the rising fortunes of the second dealer. The values obtained by elimination, or $y = 6$ when $x = 7$, show when these lines cross each other, and the distance from the axis x up to the point of intersection represents what each will then be worth.

TABLE 3.

$x.$	$y =$ Fortune of		Date.
	Dealer 1	Dealer 2.	
0	+ 20	− 15	Jan. 1
1	+ 18	− 12	" 2
2	+ 16	− 9	" 3
3	+ 14	− 6	" 4
4	+ 12	− 3	" 5
5	+ 10	0	" 6
6	+ 8	+ 3	" 7
7	+ 6	+ 6	" 8
8	+ 4	+ 9	" 9
9	+ 2	+ 12	" 10
10	0	+ 15	" 11
15	− 10	+ 30	" 16
20	− 20	+ 45	" 21

5. The area of a field is sixteen square miles. Its north and south sides have a length y, and its east

and west sides have a length x. Then by the condition imposed

$$xy = 16. \quad \ldots \quad \ldots \quad (7)$$

It is evident that either x or y may have any value between zero and an infinite value, but the other must then have a definite value such that their product is 16.

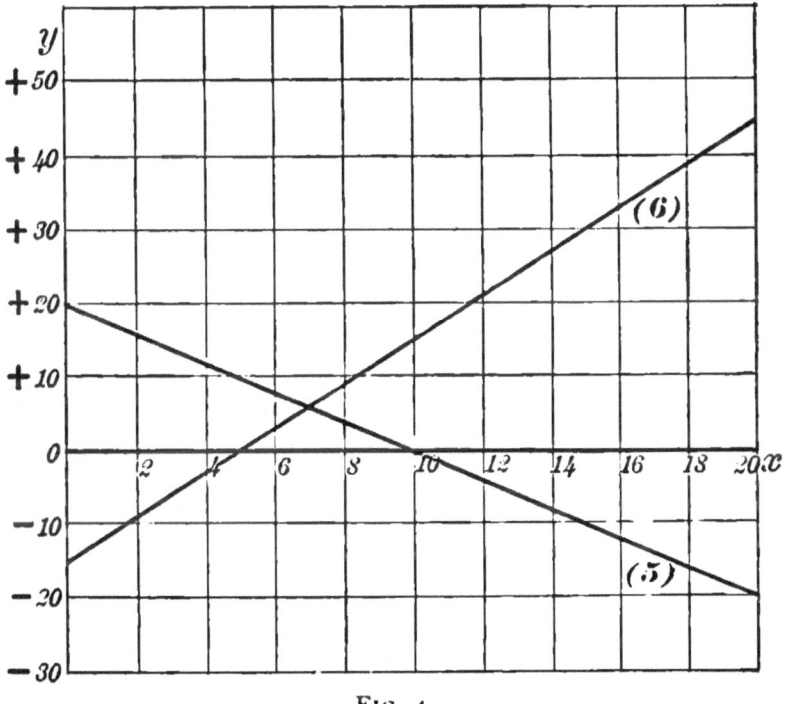

FIG. 4.

If we assume values of x, we may compute what y must be in order that their product may be 16. For convenience in making the diagram the fractional values in the table are also reduced to decimal form. It is evident that if the breadth of the field were only a thousandth of a mile, the length must be 16000 miles, and as either y or x approaches zero, the other must approach an infinite value.

TABLE 4.

x.	y.
$\frac{16}{10} = 1.60$	10
$\frac{16}{9} = 1.77$	9
$\frac{16}{8} = 2.00$	8
$\frac{16}{7} = 2.28$	7
$\frac{16}{6} = 2.66$	6
$\frac{16}{5} = 3.33$	5
$\frac{16}{4} = 4.00$	4
$\frac{16}{3} = 5.33$	3
$\frac{16}{2} = 8.00$	2
9.00	$\frac{16}{9} = 1.77$
10.00	$\frac{16}{10} = 1.60$

The values in Table 4 giving the corresponding sides of the field may be laid off on the axes y and x of Fig. 5, to the scale of miles there shown. Assuming any length to represent one side of the field, the other length is so chosen that the product xy is 16. This is in effect equivalent to assuming one corner of the field to be at the intersection of the two axes, the sides adjacent lying along the axes. The opposite corner of the field is at some point on the curve of

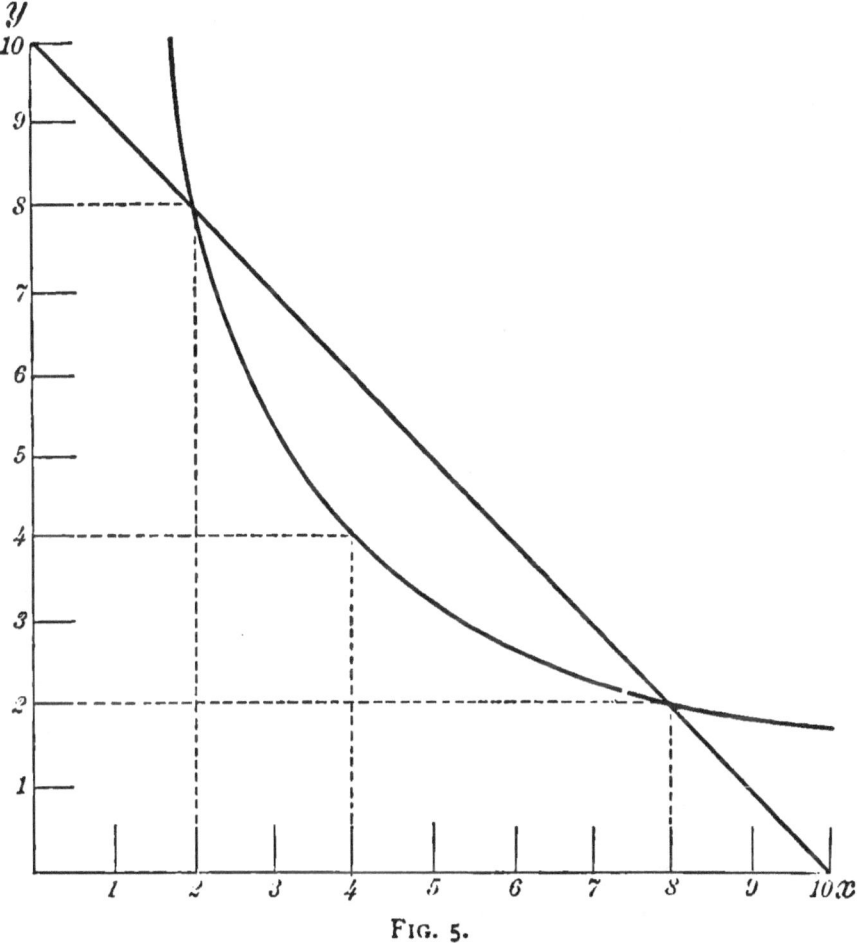

FIG. 5.

Fig. 5. The sides of the fields are in an inverse ratio. If one is doubled the other must be halved.

6. Let us assume that the sides of a field vary as before, and that the number of miles around it is 20. Between what limits may the lengths of the sides vary?

By the condition imposed, $2x + 2y = 20$, or

$$x + y = 10. \quad . \quad . \quad . \quad . \quad (8)$$

For computation of y for assumed values of x we may write the equation

$$y = 10 - x.$$

It is evident that if $x = 10$, y will equal zero. If x were assumed greater than 10, y would be negative. In this case negative values of either x or y do not admit of simple physical interpretation, and they need not be here discussed. When $x = 0$, y will be 10, or the two long sides of the field would alone require 20 miles of fencing. If x were less than zero, or negative, y would be greater than 10, and these two long sides would exceed the limiting value of 20 miles. We shall then assume values of x between 0 and 10, and shall for each case compute the area of the field so determined. See Table 5.

It is evident that if equations (7) and (8) are separately considered, we cannot say that x or y have any particular values. We have given various values to x and have computed y. The values of x and y from the last table have been plotted in Fig. 5 as in the previous table. They determine the inclined straight line.

TABLE 5.

x.	$y = 10 - x$.	Area $= xy$.
0	10	0
1	9	9
2	8	16
3	7	21
4	6	24
5	5	25
6	4	24
7	3	21
8	2	16
9	1	9
10	0	0

Let us now assume that the x and the y of equation (7) are the same as those of (8), and find x and y by elimination. The physical meaning of this is that we are to find the sides of a field which must have an area of 16 square miles, and it must also require 20 miles of fence to enclose it. We have then the two equations

$$xy = 16,$$
$$2x + 2y = 20,$$
or
$$x + y = 10.$$

Hence
$$x(10 - x) = 16,$$
$$x^2 - 10x = -16.$$

Adding 25 to both members,
$$x^2 - 10x + 25 = 9,$$
$$x - 5 = \pm 3,$$
$$x = 8 \text{ or } 2.$$

When $x = 8$, y must equal 2, and when $x = 2$, y must equal 8. These are the values corresponding to the two points of intersection of the curve and the straight line in Fig. 5.

If $2x + 2y = 16$ instead of 20, the straight line of Fig. 5 would drop two divisions, passing through divisions 8 on the axes. It would then be tangent to the curve. The values of x and y determined by elimination would then be $y = 4$, $x = 4$. This would mean that the field must be square, with its sides 4 miles long. This is the only possible case that will satisfy the condition that the area of the field is to be 16 square miles and the distance around it is to be 16 miles.

If $x + y$ is assumed less than 8, the straight line will not intersect the curve. It is impossible to have a rectangular field of 16 square miles, and so shape the field that less than 16 miles of fence will be required.

The equations will tell the same story. Take general values for the constants and write
$$xy = a,$$
$$x + y = \frac{b}{2}.$$

Eliminating y,

$$x\left(\frac{b}{2} - x\right) = a,$$

$$x^2 - \frac{b}{2}x = -a,$$

$$x^2 - \frac{b}{2}x + \frac{b^2}{16} = \frac{b^2}{16} - a.$$

$$\therefore x = \frac{b}{4} \pm \sqrt{\frac{b^2}{16} - a}.$$

In this we see that if $\frac{b^2}{16}$ is less than a, the quantity under the radical will be negative in sign, and the extraction of its square root is impossible. This would mean that an impossible condition had been assumed. In the prior example we made $a = 16$. If $\frac{b^2}{16} = a = 16$, then $b = 16$. The value of x would then be $\frac{1}{4}b = 4$.

In general we find that when the distance around any field of area a is least, $\frac{b^2}{16} = a$ or $\frac{b}{4} = \sqrt{a}$. Stated in words this means that one fourth of the distance around the field is equal to the square root of the area. This can only be true when the field is square in form.

7. Let us have given us the two equations

$$y + x = 4, \quad \ldots \ldots \quad (9)$$

$$y + x = 2. \quad \ldots \ldots \quad (10)$$

Viewing these equations from a common-sense standpoint, we should at once decide that whatever y and x may be, their sum cannot be both 4 and 2. But it is also true that their sum may be either 4 or 2. If $x = 1000$, then if y be -996 the sum will be 4, while if y be -998 it will be 2. In other words, y and x cannot have simultaneous values in the two equations.

We will asume various values of x ranging from -4 to $+6$, and compute the values of y from equations (9) and (10). These values are given in columns 2 and 3 of Table 6.

TABLE 6.

x.	y.	
	(9)	(10)
-4	$+8$	$+6$
-3	$+7$	$+5$
-2	$+6$	$+4$
-1	$+5$	$+3$
0	$+4$	$+2$
$+1$	$+3$	$+1$
$+2$	$+2$	0
$+3$	$+1$	-1
$+4$	0	-2
$+5$	-1	-3
$+6$	-2	-4

The values of x and y from (9) are laid off in a diagram in Fig. 6. They determine the upper straight line. The values of y from (10) being laid off, we get the lower straight line.

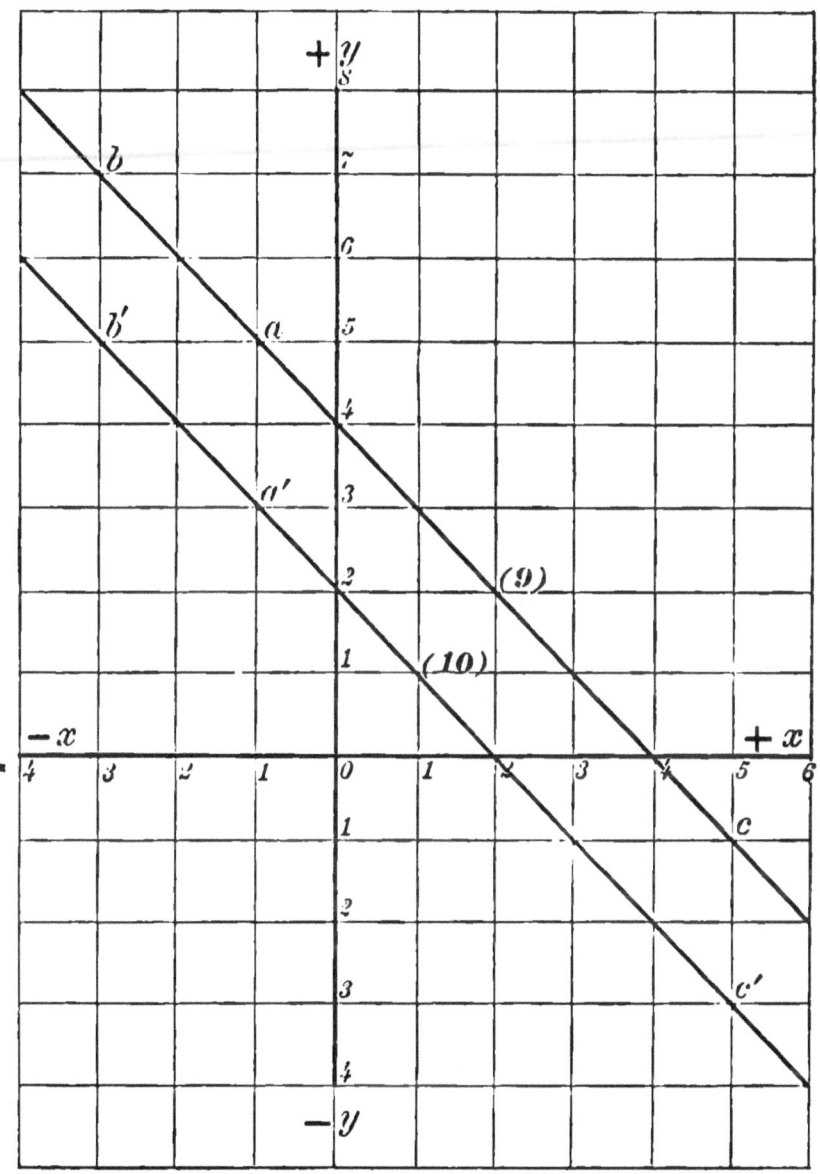

Fig. 6.

These lines are parallel to each other. There is therefore no point on the lines where, the x of one being the same as the x of the other, the y of one will be the same as the y of the other. The two points c, c' have the same value of $x = +5$, but the corresponding y of one is -1 and of the other is -3. The points a, a' have a common $x = -1$, but the values y are $+3$ and $+5$. The points a, b' have the same value $y = +5$, but the values of x are -1 and -3.

This is the geometrical explanation of the fact that when we assume that x and y are alike in (9) and (10) and seek to determine their value by elimination, we find ourselves baffled by the fact that we cannot eliminate one without eliminating the other. The equations are both true, but they are not simultaneous.

If in place of equation (10) we take either of the equations

$$x + 2y = 4, \quad \ldots \quad (11)$$
$$2x + y = 8, \quad \ldots \quad (12)$$

and combine them with (9), we get in either case the values $y = 0$ and $x = +4$. Both of the lines represented by these equations will cross the line representing equation (9) at the point where it crosses the axis x. One of them will cross the axis y at a point where $y = 8$, and the other where $y = 2$. This is easily seen by inspection of the two equations. If y

be made equal to zero, both of these equations give for x the value 4. This is also true for equation (9). If $x = 0$, $y = 4$ for (9), and for (11) and (12) when $x = 0$, $y = 2$ and 8.

8. Find x and y in the following equations:

$$x^2 + xy = 84, \quad \ldots \quad (13)$$
$$x^2 - y^2 = 24. \quad \ldots \quad (14)$$

Solving these equations for y, we have

$$y = \frac{84}{x} - x; \quad \ldots \quad (13)'$$
$$y = \pm \sqrt{x^2 - 24}. \quad \ldots \quad (14)'$$

Assuming values of x between -10 and $+10$, we have the values of y in Table 7.

In equation (14) the value of y becomes zero when $x^2 = 24$ or $x = 4.899$. The sign $\pm$ indicates that for each value of x^2 there are two equal values of y with unlike sign. The axis x is therefore a line of symmetry. When x^2 is less than 24, the value of y becomes imaginary. The value of y is also the same whether x is positive or negative in sign, since x is involved to the second power. Hence between $x = +\sqrt{24}$ and $-\sqrt{24}$ the equation does not give real values. Each of the curves obtained by laying off the values of y for (13) and for (14), with the values of x given in the table, consists of two isolated branches as is shown in Fig. 7. They intersect at two points $x = +7$ and $y = +5$, and $x = -7$ and $y = -5$. These are the values that are found by making the equations simultaneous and solving by elimination.

TABLE 7.

x.	Value of y.	
	(13)	(14)
-10	$+ 1.60$	± 8.72
$- 9$	$- 0.33$	± 7.55
$- 8$	$- 2.50$	± 6.32
$- 7$	$- 5.00$	± 5.00
$- 6$	$- 8.00$	± 3.46
$- 5$	$- 11.80$	± 1.00
$- 4$	$- 17.00$	
$- 3$	$- 25.00$	
$- 2$	$- 40.00$	
$- 1$	$- 83.00$	
0	∞	
$+ 1$	$+ 83.00$	
$+ 2$	$+ 40.00$	
$+ 3$	$+ 25.00$	
$+ 4$	$+ 17.00$	
$+ 5$	$+ 11.80$	± 1.00
$+ 6$	$+ 8.00$	± 3.46
$+ 7$	$+ 5.00$	± 5.00
$+ 8$	$+ 2.50$	± 6.32
$+ 9$	$+ 0.33$	± 7.55
$+ 10$	$- 1.60$	± 8.72

9. Given the equation
$$400y^2 + 9x^2 - 200xy + 64 = 0. \quad (15)$$
This equation when solved for y becomes
$$y = \frac{x}{4} \pm \frac{1}{5} \sqrt{x^2 - 4}. \quad (15)'$$

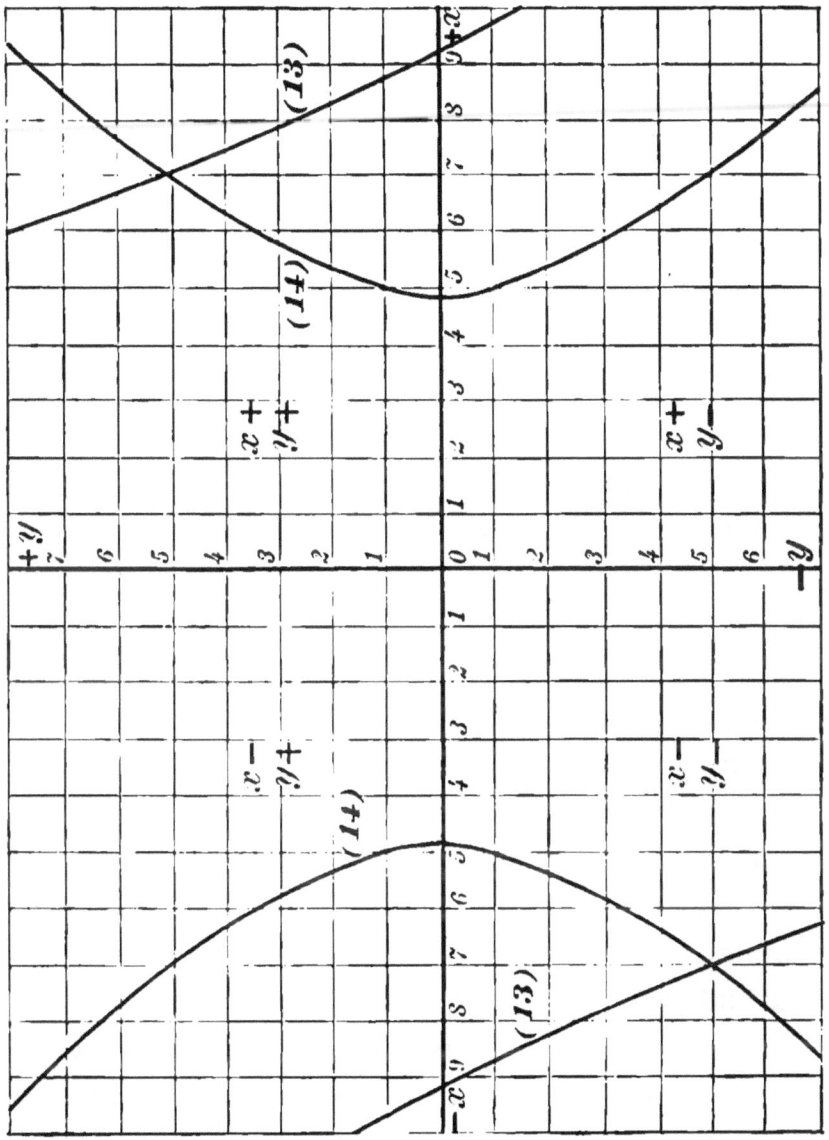

Fig. 7.

In this form the equation shows that the value y is made up of two parts, viz., the term $\frac{1}{4}x$ and a term having a $\pm$ sign. Both terms increase with x. Each value of y will therefore have two values for each value of x. The mean of those two values will be $\frac{1}{4}x$, since one of these values is $\frac{1}{4}x$ increased by the $\pm$ term, and the other is $\frac{1}{4}x$ diminished by an equal value. This at once shows that the line whose equation is

$$y = \frac{x}{4}$$

is a line of symmetry for the curve whose equation is (15)'. This line bisects the chords of the curve drawn parallel to the axis y. It is the line marked a, b in Fig. 8. The two terms of (15)' are separately computed in Table 8. For values of x less than 2 the $\pm$ term is imaginary.

The two values of y for each value of x are plotted in Fig. 8.

It is evident that if either equation (13) or (14) be combined with equation (15) there will be four values of x and four of y. This is apparent from an inspection of the curves in Figs. 7 and 8. Each branch of (14) or (13) will intersect the corresponding branch (15) at two points. Two values of x and y will therefore be positive and two will be negative.

10. From what has preceded it will have been

TABLE 8.

x.	$\frac{x}{4} \pm \frac{1}{5}\sqrt{x^2-4}$.	y.		
+ 10	+ 2.50 ± 1.96	+ 4.46	or	+ 0.54
+ 9	+ 2.25 ± 1.76	+ 4.01	"	+ 0.49
+ 8	+ 2.00 ± 1.55	+ 3.55	"	+ 0.45
+ 7	+ 1.75 ± 1.34	+ 3.09	"	+ 0.41
+ 6	+ 1.50 ± 1.13	+ 2.63	"	+ 0.37
+ 5	+ 1.25 ± 0.92	+ 2.17	"	+ 0.33
+ 4	+ 1.00 ± 0.69	+ 1.69	"	+ 0.31
+ 3	+ 0.75 ± 0.45	+ 1.20	"	+ 0.30
+ 2.5	+ 0.625 ± 0.30	+ 0.925	"	+ 0.325
+ 2	+ 0.50 ± 0.00	+ 0.50	"	+ 0.50
+ 1				
0				
− 1				
− 2	− 0.50 ± 0.00	− 0.50	"	− 0.50
− 2.5	− 0.625 ± 0.30	− 0.925	"	− 0.325
− 3	− 0.75 ± 0.45	− 1.20	"	− 0.30
− 4	− 1.00 ± 0.69	− 1.69	"	− 0.31
− 5	− 1.25 ± 0.92	− 2.17	"	− 0.33
− 6	− 1.50 ± 1.13	− 2.63	"	− 0.37
− 7	− 1.75 ± 1.34	− 3.09	"	− 0.41
− 8	− 2.00 ± 1.55	− 3.55	"	− 0.45
− 9	− 2.25 ± 1.76	− 4.01	"	− 0.49
− 10	− 2.50 ± 1.96	− 4.46	"	− 0.54

observed that an equation of the first degree between two variables x and y may be represented by a straight line. To draw this line only two points need be determined. The most convenient points are the intersections of the line with the axes of y and x. Thus in the equation

$$x + 2y = 4, \quad \ldots \quad (16)$$

if y be made zero we have $x = 4$. This line crosses the x-axis at a point when $x = 4$. Making $x = 0$, we have $y = +2$. This line therefore crosses the

GRAPHICAL ALGEBRA. 31

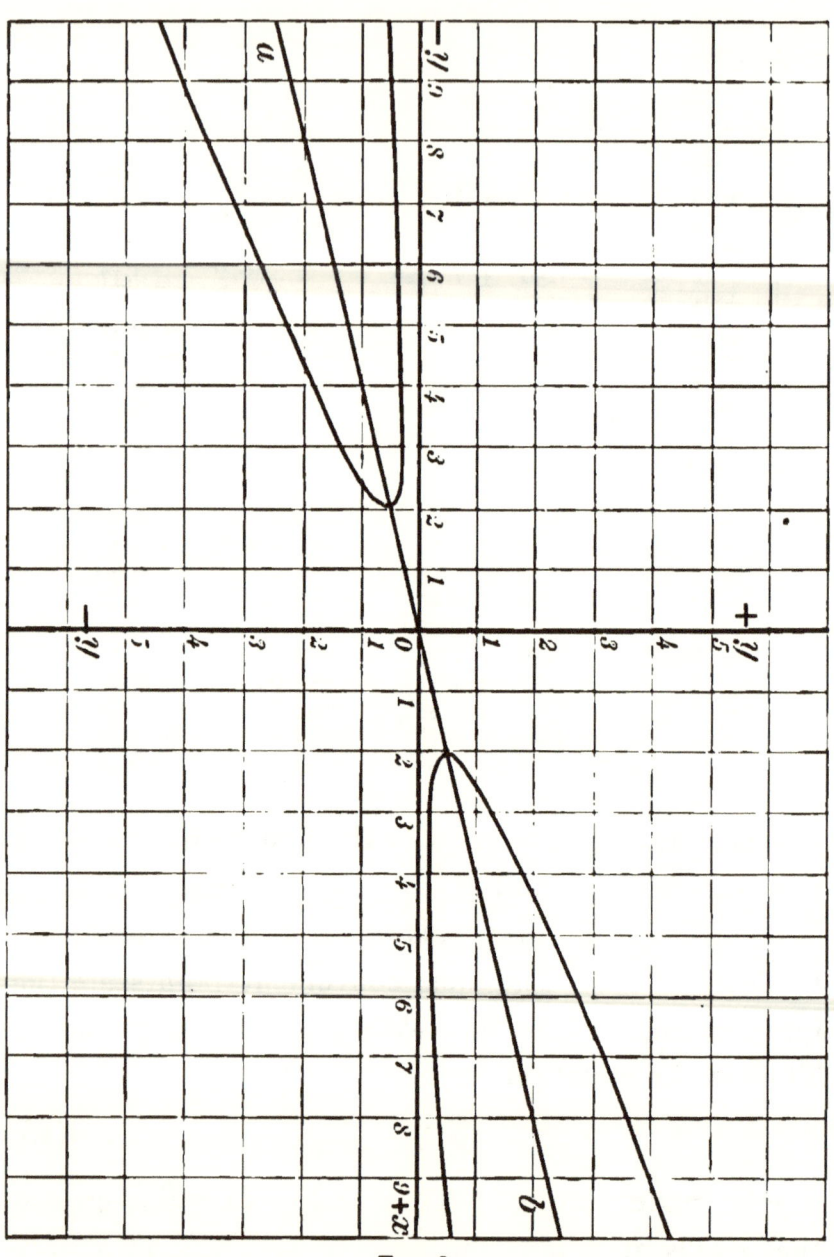

Fig. 8.

axis y, where $y = +2$. This line is shown in Fig. 9, where it is marked (16). In the equation

$$x - 2y = 4 \quad . \quad . \quad . \quad . \quad . \quad (17)$$

the intersection on the axis of y is -2 instead of $+2$. In the equation

$$-2x + y = 4 \quad . \quad . \quad . \quad . \quad . \quad (18)$$

the line crosses the axes y and x at points $y = +4$ and $x = -2$. In the equation

$$-2x - y = 4 \quad . \quad . \quad . \quad . \quad . \quad (19)$$

the intersections are $y = -4$ and $x = -2$. All of these lines are shown in Fig. 9.

When the equation is of a degree higher than the first, the line representing it is in general curved. The intersections with the axes, if such intersections exist, may be found in the same way as in the straight line, but the position of other points of the line must be computed. At places where such lines bend sharply the points so determined must be closer together than at points where there is little curvature.

11. When an equation contains three unknown or variable quantities, x, y, and z, three axes of reference are used in order to represent the relation. Such a method is in general use in such a case as the location of the summit of a mountain. We should give the

latitude and longitude of the place, and its altitude above any assumed datum plane, such as the level of the sea.

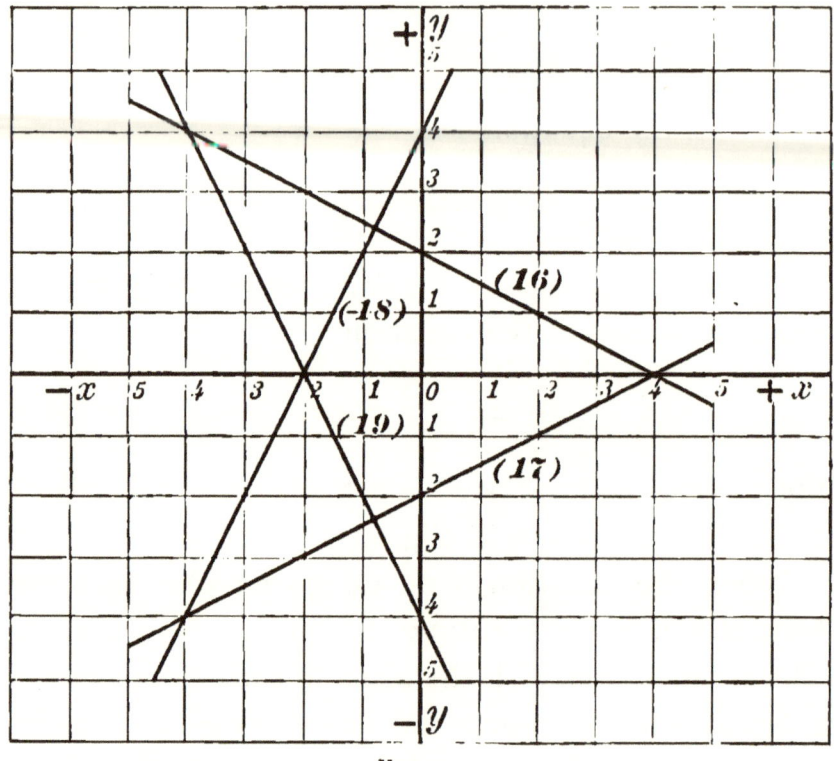

Fig. 9.

Fig. 10 shows a system of three reference planes, intersecting each other at right angles, in three axes. The three axes intersect in a common point called the origin. Measured from the origin, a distance to the right along the axis x is $+x$, and to the left is $-x$. Distances upward and towards the observer are respectively $+z$ and $+y$, while distances in the opposite directions have the $-$ sign.

Fig. 11 shows one of the eight trihedral angles of Fig. 10. It is intersected by a plane whose intersections with the three reference planes make angles of

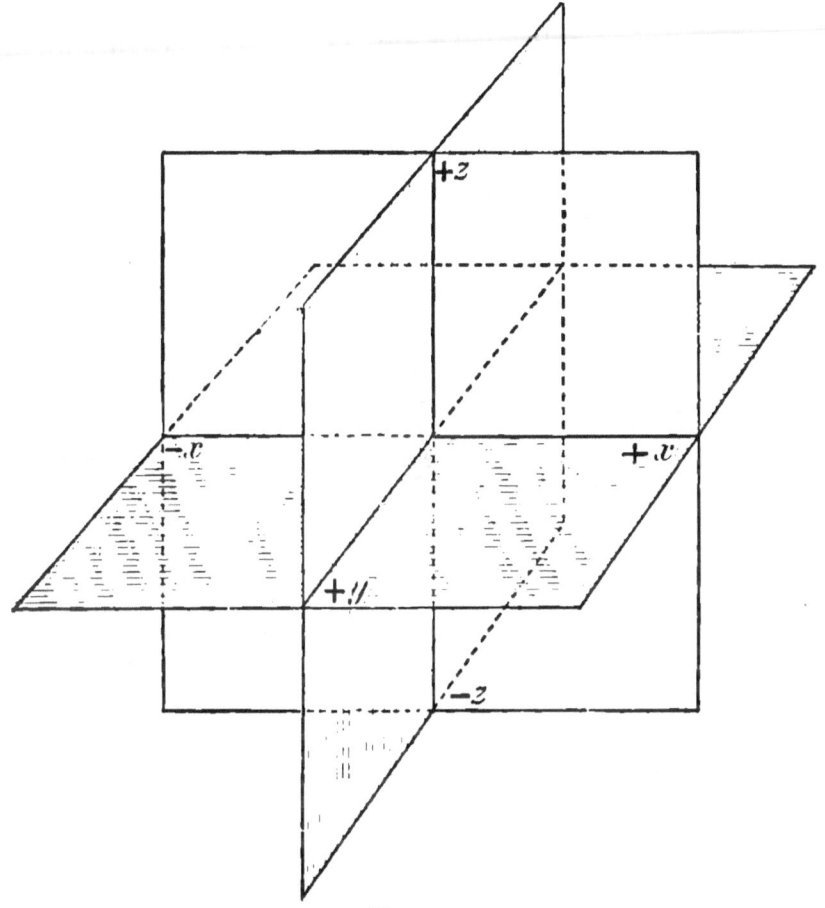

FIG. 10.

45° with the three axes. The axes are each intersected at a distance 8 from the origin. The intersection of the inclined plane with the reference plane x, z has for its equation

$$x + z = 8.$$

This line extends infinitely in both directions, but only the part between the axes z and x is shown in

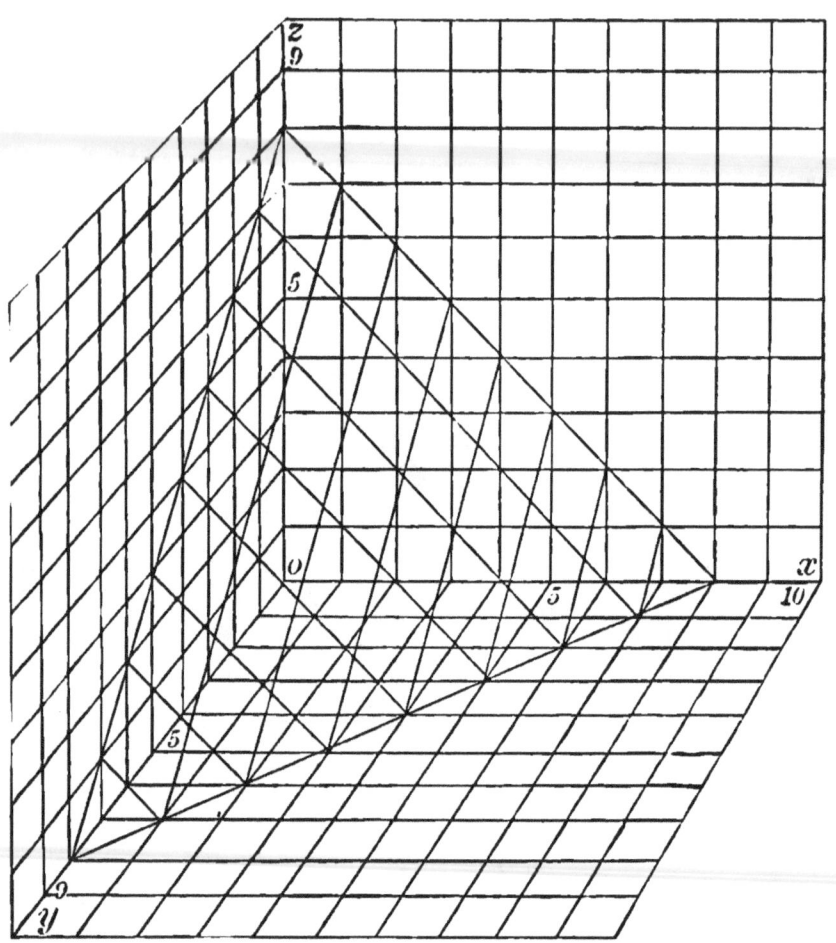

Fig. 11.

the figure. In front of this line a distance $y = 1$ from the back plane of x, z is a line in the inclined plane whose intersections with the bottom and left reference planes are distant $8-1$ or $8-y$ from the axis y. The equation of this line is evidently

$$x+z = 8-1 = 8-y.$$

Similarly the next line in the figure, at a distance $y = 2$ from the back plane, has the equation

$$x+z = 8-2 = 8-y.$$

From this it is apparent that the equation of any line of this series distant y from the back plane is

$$x+z = 8-y.$$

Evidently this applies equally to any line intermediate between those drawn in the figure. And the equation therefore represents the location of any point on any such line, or, in other words, any point on the plane. The equation may be written

$$x+y+z = 8. \quad . \quad . \quad . \quad . \quad (20)$$

If one starts from the origin and travels along the axis x a distance 2, and then travels parallel to the axis y a distance 6, the distance upward to the plane will be zero. If the last distance parallel to y is 5, the distance up to the plane will be 1. If the distance along y be made 7, the distance to the plane parallel to the axis z will be -1. In all cases, whatever may be the lengths of the paths along x, y, and z, if the sum of these distances is $+8$, the journey will terminate at a point on the inclined plane. For this reason equation (20) is said to be the equation of the plane.

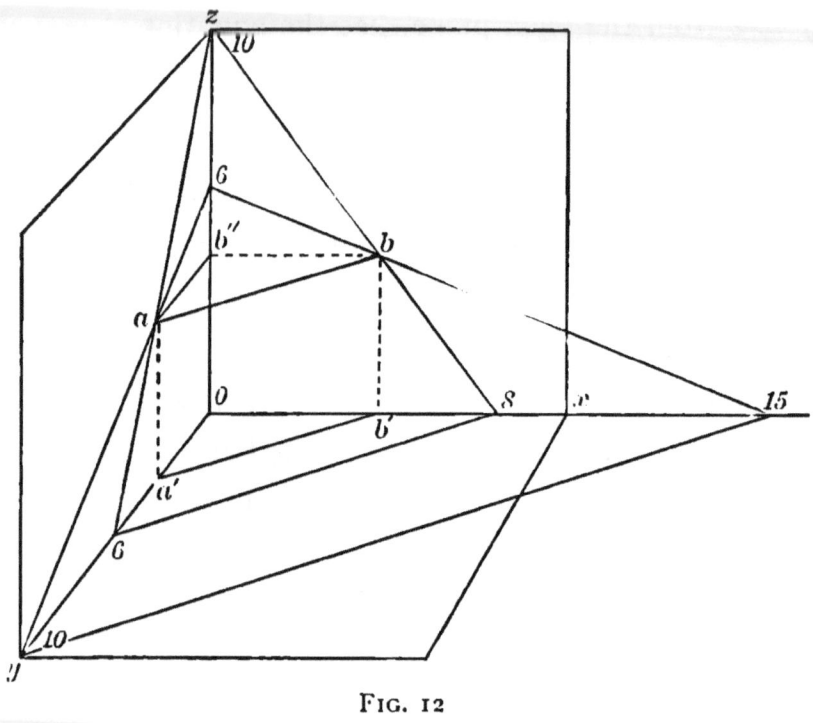

Fig. 12

Fig. 12 shows the same reference planes, with two planes intersecting them. The equations of these two planes are

$$2x + 3y + 5z = 30, \quad \ldots \quad (21)$$

$$15x + 20y + 12z = 120. \quad \ldots \quad (22)$$

That these are the equations of these two planes may be verified as follows: The equations are to be true for all values of x, y, and z. They will therefore be true when $y = 0$. The two equations then become

$$2x + 5z = 30, \quad \ldots \quad (23)$$

$$15x + 12z = 120. \quad \ldots \quad (24)$$

These are the equations of the two intersecting lines on the back plane of x, z. These equations hold for each point on those lines. They hold, therefore, when $x = 0$. In that case we find from the first equation $5z = 30$, or $z = 6$. This determines the point where this plane cuts the axis z. For the other plane, $12z = 120$, or $z = 10$.

In a similar way if $z = 0$, in equations (23) and (24), we have for the intersections on the axis of x, $2x = 30$ or $x = 15$, and $15x = 120$ or $x = 8$.

Likewise if x and z are made zero in (21) and (22) the intersections on y are determined by the resulting equations. We have $3y = 30$ or $y = 10$, and $20y = 120$ or $y = 6$.

By elimination of x or y in (23) and (24) we may find values for those quantities. They are $z = \frac{210}{51} = 4.12$ and $x = \frac{240}{51} = 4.71$. But these values of x and z, which determine the point b of Fig. 12, were obtained on the assumption that $y = 0$. By giving y other values, other points on the line a, b, or a, b

produced, may be obtained. For example, if x be made zero in (21) and (22), we have

$$3y + 5z = 30, \quad \ldots \quad (21)'$$
$$5y + 3z = 30. \quad \ldots \quad (22)'$$

By elimination we find $y = z = \tfrac{15}{4} = 3.75$. In fact, if we assume x, y, and z to mean the same in (21) and (22), which are the equations of the two planes, then the values x, y, and z can only refer to the points which are common to the two planes, and which lie in their line of intersection a, b. This is the geometrical meaning of the fact that the three unknown quantities cannot be determined with only two equations.

If we combine (21) and (22) by the elimination of x, we have

$$5y + 51z = 210. \quad \ldots \quad (25)$$

This is the equation of a straight line on the plane y, z. It is the relation between the distances y and z of points along the line a, b. If the points of this line be projected parallel to x upon the line a, b'', equation (25) will be the equation of that projection. For each point in a, b will have the same y and z that its projection has.

This projection will therefore intersect the axis z (y being then zero in (25)), where $z = \tfrac{210}{51}$. This is the value previously found for b, b'. Making $z = 0$

in (25) we find that this projection intersects the axis y, where $y = 42$. If y be made $\tfrac{15}{4}$ in (25), we find z to be $\tfrac{15}{4}$ also, which was previously determined.

Eliminating z in (21) and (22), we have

$$51x + 64y = 240. \quad . \quad . \quad . \quad . \quad (26)$$

This is the equation of the line a', b', which is the projection of a, b on the plane x, y. Making $y = 0$ in this equation, the distance to the intersection with the axis x at b' is $x = \tfrac{240}{51}$, as was previously found.

Now it is evident that if we pass a third plane through this trihedral angle not parallel to either of the two planes of Fig. 12, it will intersect each of them in two additional lines like a, b. These three lines of intersection will intersect each other in a common point, as the intersections of the reference planes, or the axes x, y, and z, intersect in a common point, the origin. This point will be the only point which will lie in all three of the planes. That is the only point for which x, y, and z can have the same value in the three equations. The values of x, y, and z obtained by elimination will locate this point of intersection.

12. Let us assume the equations of three planes as follows:

$$2x + 4y + 4z = 20, \quad . \quad . \quad . \quad (27)$$
$$5x + 2y + 5z = 20, \quad . \quad . \quad . \quad (28)$$
$$4x + 5y + 2z = 20. \quad . \quad . \quad . \quad (29)$$

There is a line of intersection between planes (27) and (28), another between planes (27) and (29), and a third between planes (28) and (29). Combining these equations in that order, if we eliminate y we shall obtain the equation of the projections of those three lines on the plane z, x. In this way we have

$$4x + 3z = 10, \quad \ldots \quad (30)$$
$$-3x + 6z = 10, \quad \ldots \quad (31)$$
$$17x + 21z = 60. \quad \ldots \quad (32)$$

Since the lines in which the planes intersect each other have a common point of intersection, their projections, represented by the last equations, must have a common point. Therefore any two of those equations will determine the value of x and z for that point. By elimination with either (30) and (31), (30) and (32), or (31) and (32), we find $x = \frac{30}{33}$, $z = \frac{70}{33}$.

If these values of x and z be substituted in the equations of either one of the three planes, (27), (28), or (29), they will each give the value $y = \frac{80}{33}$. This shows that the point whose co-ordinates are $x = \frac{30}{33}$, $y = \frac{80}{33}$, $z = \frac{70}{33}$ is a point common to the three planes.

The three planes with their intersections are shown in Fig. 13.

If the coefficient of x in (27) were -2 instead of $+2$, the intersection of that plane with the x-axis would be at a point -10 instead of $+10$. The points of intersection of that plane with the axes z

and y would be unchanged. The triangle determined by the intersections with the three axes would be in a symmetrically reversed position in the trihedral angle

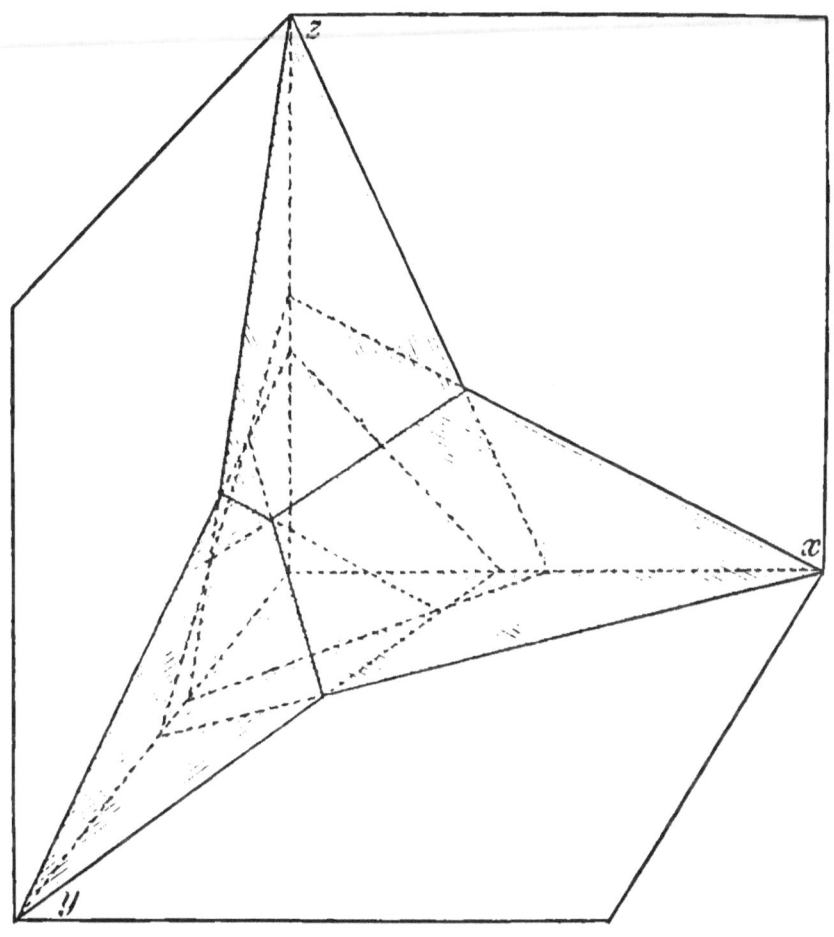

Fig. 13.

to the left of the one shown in Fig. 13. (See Fig. 10.) This would change the position of two of the lines of intersection between the three planes, but would not change the other.

GRAPHICAL ALGEBRA. 43

13. Instead of eliminating y in equations (27), (28), and (29), as in the last article, we may reach precisely the same result by eliminating either x or z. In either case there will result three equations like (30), (31), and (32), involving the other two quantities, any two of which will determine those two quantities. These quantities may then be put into any one of the original equations, and the values $x = \frac{30}{33}$, $y = \frac{80}{33}$, $z = \frac{70}{33}$ will be determined. This solution is merely a projection of the lines of intersection on the other two reference planes.

If the signs of the coefficients in the original equations are varied, the point common to the three planes may be thrown into any one of the eight trihedral angles shown in Fig. 10. If x in the final solution is $+$, the common point will lie in one of the four trihedral angles to the right of the plane y, z of Fig. 10. If y is also positive, it will be in one of the two to the right of y, z, and in front of x, z. If z is also negative, it will lie in the lower angle of these two.

14. An equation of the second degree between three variables is represented by a curved surface.

Assume the equation

$$zx + 2y = 0 \quad \ldots \quad \ldots \quad (33)$$

This surface may be represented by lines representing sections of the surface, as was done in case of the plane in Fig. 11. For example, make $x = 10$ in the last equation. It then becomes, on solving for z,

$$z = -\tfrac{1}{2}y.$$

This is the equation of a straight line. In Fig. 14 this line is the extreme line to the right of the model,

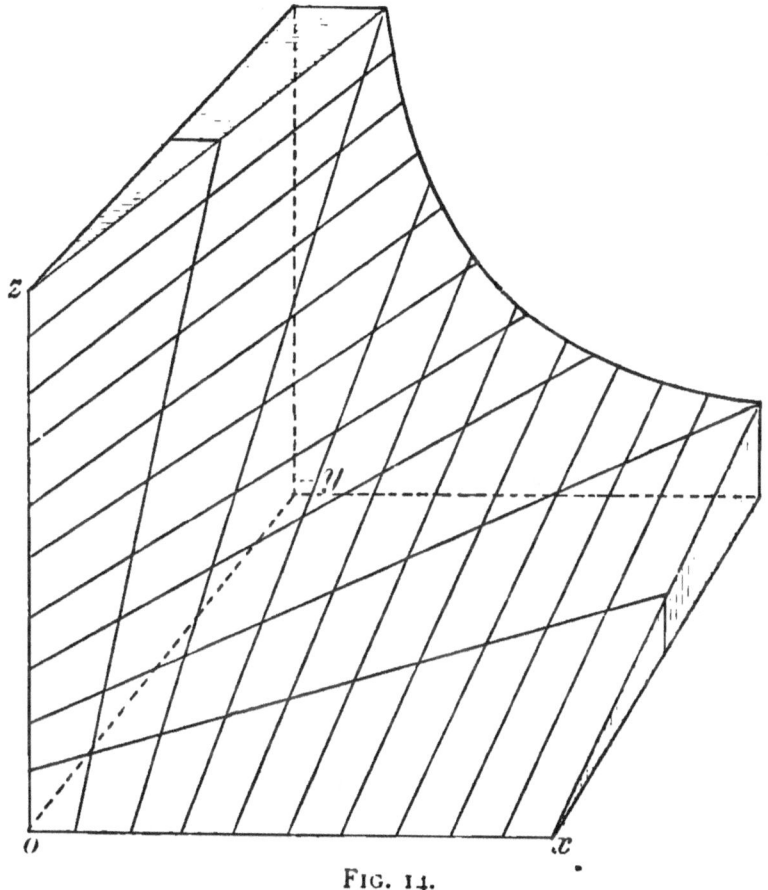

Fig. 14.

where $x = 10$. From the front edge of the model to the rear section y varies from 0 to -10. The values of z in this model are all positive, the value at the rear right-hand corner, where $x = 10$ and $y = -10$, being $+ 2$.

The straight line adjoining the previous one, and along which $x = 9$, is represented by the equation

$$z = -\tfrac{2}{9}y.$$

The line next to the left has for its equation

$$z = -\tfrac{2}{8}y = -\tfrac{1}{4}y.$$

The final line of this series shown in the figure, where $x = 1$, has for its equation

$$z = -2y.$$

Each of these lines represents an intersection with the surface whose equation is (33) by a vertical plane at right angles to the axis x. For all points of this plane x is the same. Making x constant and assigning various constant values to it enables us to compute the curves representing the sections of the surface at any distance x from the reference plane z, y.

The other series of sections shown in Fig. 14 as crossing lines of the former series are characterized by the condition $z =$ constant.

For the uppermost of these, where $z = 10$,

$$x = -\tfrac{2}{10}y = -\tfrac{1}{5}y.$$

For the lowermost one, where $z = 1$,

$$x = -2y.$$

These lines correspond to intersections with the surface made by a horizontal plane cutting the vertical axis at any distance z from the origin.

The curved line at the back of the model, where $y = -10$, has for its equation

$$zx = +20.$$

For the upper point of this curve, where the model is cut off on a horizontal section, $x = 2$ and $z = 10$, while for the lowest point shown $x = 10$ and $z = 2$. In both cases, and for any other point on this curve, $zx = 20$.

Along a similar section midway between the front and rear of the model $y = -5$. The equation of such section would be

$$zx = 10.$$

Along the front edge of the model $y = 0$, and along this plane, as (33) shows, the product $zx = 0$.

All of these equations are special cases of equation (33). At any point on the horizontal reference plane, determined by x and y, the vertical distance z to the surface may be found from (33). This value is

$$z = -2\frac{y}{x}.$$

This surface extends into other angles formed by the reference planes of Fig. 10, but it is not thought desirable to go further into the subject here.

15. There are many surfaces in which all of the sections at right angles to the axes are curved. In Fig. 14 those sections at right angles to the axes z and x are not curved.

Assume the equation

$$4xz - 3yz + 6xy = 0. \quad . \quad . \quad . \quad (34)$$

If we solve this equation for z, we have

$$z = -\frac{6xy}{4x - 3y}. \quad . \quad . \quad . \quad (35)$$

We will examine this surface in the reference angle where x is positive and y negative in sign, as in the case of Fig. 14.

If in this equation we make $x = 10$, (35) or (34) will show the relation between z and y, along a section of the surface, at a distance $x = 10$ from the plane z, y. The equation then becomes

$$z = -\frac{60y}{40 - 3y}.$$

Giving y consecutive values between 0 and -10, the values of z are computed as in Table 9 adjoining. It is evident that z increases more slowly as y becomes numerically large. If $y = -10000$, the value of z is only 19.97. As y becomes very large the value of 40 in the denominator becomes insignificant compared with $3y$. We may under these circumstances neglect the term 40. As y increases more and more, z approaches the value

TABLE 9.

y.	z.
0	0.00
− 1	+ 1.39
− 2	+ 2.61
− 3	+ 3.67
− 4	+ 4.61
− 5	+ 5.45
− 6	+ 6.21
− 7	+ 6.89
− 8	+ 7.50
− 9	+ 8.06
− 10	+ 8.57

$$z = -\frac{60y}{-3y} = +20.$$

This would be the value of z when y is negative and infinite.

The values of y and z of the table determine the section shown on the right of Fig. 15, where $x = +10$.

If $x = 5$, equation (35) becomes

$$z = -\frac{30y}{20 - 3y}.$$

If consecutive values from 0 to -10 be substituted in this equation, we may construct another table like the previous one. The two tables with others corresponding to intermediate values of x may be plotted or drawn to the same scale, and the curves may be

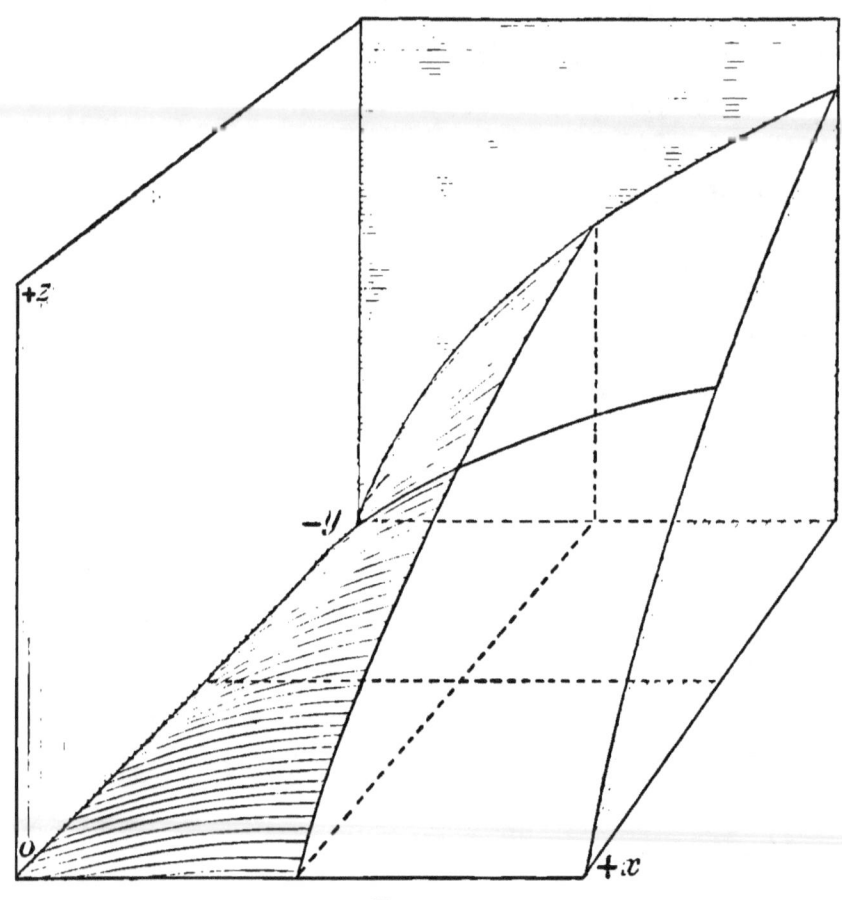

FIG. 15.

cut out in paper, cardboard, or zinc sheet, and fastened in vertical position relatively to each other as determined by their respective values of x, as is shown in Fig. 15.

If $y = -10$ in (35), that equation becomes

$$z = \frac{60x}{4x + 30}.$$

If we now give to x consecutive values from 0 to $+10$, we shall obtain the values of x and z in Table 10, which will enable us to construct the curve forming the back side of the model in Fig. 15. The last values in the Tables 9 and 10 are identical, and they correspond to the corner of the model to the rear and right.

TABLE 10.

x.	z.
0	0.00
1	+ 1.76
2	+ 3.16
3	+ 4.28
4	+ 5.21
5	+ 6.00
6	+ 6.67
7	+ 7.24
8	+ 7.74
9	+ 8.18
10	+ 8.57

If (33) and (34) are combined with each other by the elimination of z, we have

$$\frac{6xy}{4x - 3y} = \frac{2y}{x}.$$

If $y = 0$, this equation is satisfied for any value of x. The equation will then read $0 = 0$. If x may have any value when $y = 0$, then (33) and (34) with the condition $y = 0$ must be true for any value of x. These equations both reduce to the form $zx = 0$. Since this must be true for any value of x, it follows that z must be zero when $y = 0$, and that x may then have any value. If we refer to Figs. 14 and 15, we find that the axis x lies in each of these surfaces. The surfaces therefore intersect each other along that line, the condition of this intersection being $z = 0$, $y = 0$, $x =$ any value.

The last equation must also be satisfied when the common factor y is stricken out, or when

$$\frac{6x}{4x - 3y} = \frac{2}{x}.$$

Solving this equation for x, we have

$$x = \tfrac{2}{3} \pm \sqrt{\tfrac{4}{9} - y}. \quad \ldots \quad (36)$$

This equation represents a curved line on the horizontal reference plane, directly under another intersection line of the two surfaces. By inspection of the last equation it is evident that when $y = +\tfrac{4}{9}$ the $\pm$ term becomes zero, and in that case $x = \tfrac{2}{3}$. If y be made greater than $+\tfrac{4}{9}$, the $\pm$ term becomes imaginary. When $y = 0$, x becomes $\tfrac{2}{3} \pm \tfrac{2}{3} = \tfrac{4}{3}$ or 0. This intersection line therefore crosses the axis x, the other intersection line, at the two points $x = 0$, $x = \tfrac{4}{3}$, y and z being then both zero. (See Fig. 16.) For all negative values of y there are two real values of x, symmetrically related to the value $+\tfrac{2}{3}$. The curve representing these values is shown in Fig. 16. The line of symmetry for which $x = \tfrac{2}{3}$ is the line a, b in that figure. The point where the curve crosses this line is the point where the $\pm$ term in (36) is zero, and where there is but one value of x. The values of x for various values of y computed from (36) are given in Table 11.

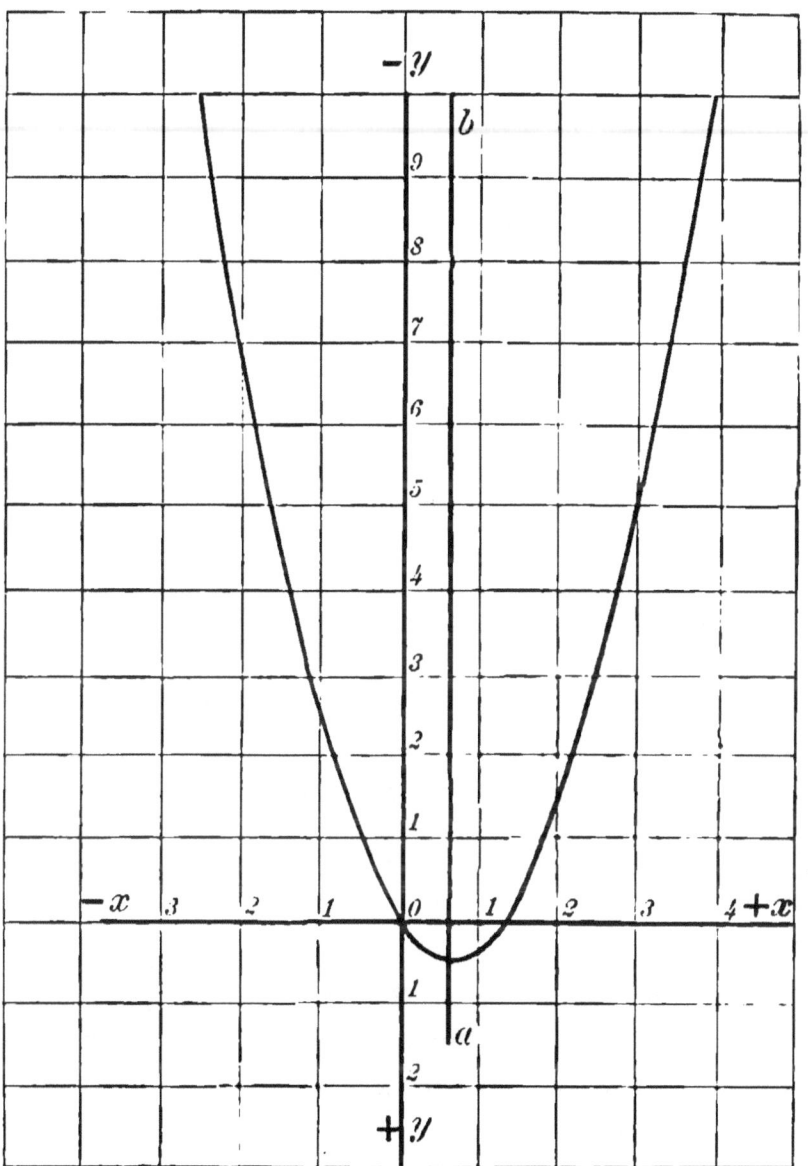

Fig. 16.

It will be of interest to examine the vertical sections of the two surfaces along the line a, b of Fig. 16.

In order to do this it is only necessary to make $x = \frac{2}{8}$ in equations (33) and (34) and compute z for each surface for various values of y between $+ \frac{6}{9}$ and -1.

These equations become

TABLE 11.

y.	z.	
$+ \frac{4}{9}$	0.667	
0	0.000	or $+ 1.333$
$- 1$	$- 0.535$	" $+ 1.868$
$- 2$	$- 0.897$	" $+ 2.230$
$- 3$	$- 1.189$	" $+ 2.522$
$- 4$	$- 1.442$	" $+ 2.775$
$- 5$	$- 1.666$	" $+ 3.000$
$- 6$	$- 1.872$	" $+ 3.205$
$- 7$	$- 2.062$	" $+ 3.395$
$- 8$	$- 2.239$	" $+ 3.572$
$- 9$	$- 2.407$	" $+ 3.740$
$- 10$	$- 2.565$	" $+ 3.898$

$$z = -3y, \quad \ldots \quad (33)'$$

$$z = -\frac{12y}{8 - 9y}. \quad \ldots \quad (34)'$$

Making $y = +\frac{4}{9}$ in these two equations, we find that $z = -\frac{4}{3}$ in both. This is the point where the curve crosses the line a, b. The values for a few points along this line are given in Table 12 adjoining. The pupil should extend these computations and draw the two section lines on the same sheet. Compute the values between $y = +10$ and $y = -10$. In a similar way other and parallel sections of the two surfaces may be computed by making $x = 2$, 3, or 4, or -2, -3, -4, etc.,

TABLE 12.

y.	z.	
	$(33)'$	$(34)'$
$+\frac{6}{9}$	$- 2.00$	$- 4.00$
$+\frac{4}{9}$	$- 1.33$	$- 1.33$
$+\frac{2}{9}$	$- 0.66$	$- 0.44$
0	0.00	0.00
$- 1$	$+ 3.00$	$+ 0.71$

in (33) and (34), and then giving y various values between $+$ 10 and $-$ 10. These sections will all cross where $y = 0$, since z for both surfaces will be zero. They will also always cross directly above the curved line of Fig. 16, the two values of z being there equal. As will be seen by (35), in which (34) is solved for z, when x and y are positive, and $x = \frac{3}{4}y$, the value of z will become infinite. If $4x > 3y$, z will be negative, and if $4x < 3y$, it will be positive.

16. We may now assume the equation of a plane, and combine it with (33) and (34). The three equations are

$$zx + 2y = 0, \quad \cdots \quad (33)$$

$$4xz - 3yz + 6xy = 0, \quad \cdots \quad (34)$$

$$x - y + z = 10. \quad \cdots \quad (37)$$

We have already eliminated z in (33) and (34) and found that the intersection projected on the plane y, x is the curve in Fig. 16 and the axis x. The equation of the curve we have found to be

$$x = \tfrac{2}{3} \pm \sqrt{\tfrac{4}{9} - y}. \quad \cdots \quad (36)$$

We are to find points common to the three surfaces. These are the points where the intersection of (33) and (34) pierces plane (37). Either (33) or (34) may be combined with (37) by eliminating z, the resulting equation being combined with (36).

Eliminating z in (33) and (37),

$$x^2 - (10 + y)x = 2y. \quad \cdots \quad (38)$$

For algebraic solution it is simplest to eliminate y in the three original equations and then find x and z in the two simplest equations resulting. For graphical discussion it is more instructive to solve the last equation for x. The result is

$$x = 5 + \tfrac{1}{2}y \pm \tfrac{1}{2}\sqrt{y^2 + 28y + 100}. \quad (39)$$

This expression for x is composed of two parts, which may be indicated by

$$x_1 = 5 + \tfrac{1}{2}y,$$
$$x_2 = \pm \tfrac{1}{2}\sqrt{y^2 + 28y + 100}.$$

The part of x represented by x_1 is represented geometrically by a straight line, which crosses the axis y, ($x_1 = 0$,) where $y = -10$, and which crosses the axis x, ($y = 0$,) where $x_1 = 5$. This line is the line c, d in Fig. 17. For the total value of x at any point we must add to the values of x for this line the $\pm$ term x_2. This $\pm$ term becomes zero when

$$y^2 + 28y + 100 = 0,$$

or when $y = -4.202$ or -23.798. For values of y intermediate between these two the $\pm$ term x_2 becomes imaginary. For values algebraically less than -23.798 (numerically greater) or greater than -4.202, the $\pm$ term is real. The values of x for assumed values of y are given in Table 13 as computed from (39).

TABLE 13.

y.	x.			
+ 4	+ 14.550	or	−	0.550
+ 1	+ 11.178	"	−	0.178
0	+ 10.000	"		0.000
− 1	+ 8.772	"	+	0.222
− 2	+ 7.464	"	+	0.536
− 3	+ 6.000	"	+	1.000
− 4	+ 4.000	"	+	2.000
− 4.202	+ 2.894			
− 23.798	− 6.899			
− 24	− 6.000	or	−	8.000
− 25	− 5.000	"	−	10.000
− 28	− 4.000	"	−	14.000
− 30	− 3.676	"	−	16.325

These values of y plotted with the double values of x determine the two isolated branches of the curve marked (33), (37) in Fig. 17. The line c, d is a line of symmetry with respect to these branches. It bisects any chord parallel to the axis x. Where this line cuts the two branches there is but one value of x. The $\pm$ term is zero.

By eliminating as before described, the two values of x are found to be $x = +\ 2.8219$ and $x = -\ 4.4886$. These are the values of x corresponding to the intersection of the two curves which these equations represent. They are both drawn in Fig. 17, one of them being reproduced from Fig. 16. These values of x in (36) or (38) give $y = -\ 4.2009$ and $y = -\ 26.1324$. The values $x = +\ 2.8219$, $y = -\ 4.2009$ locate the

GRAPHICAL ALGEBRA.

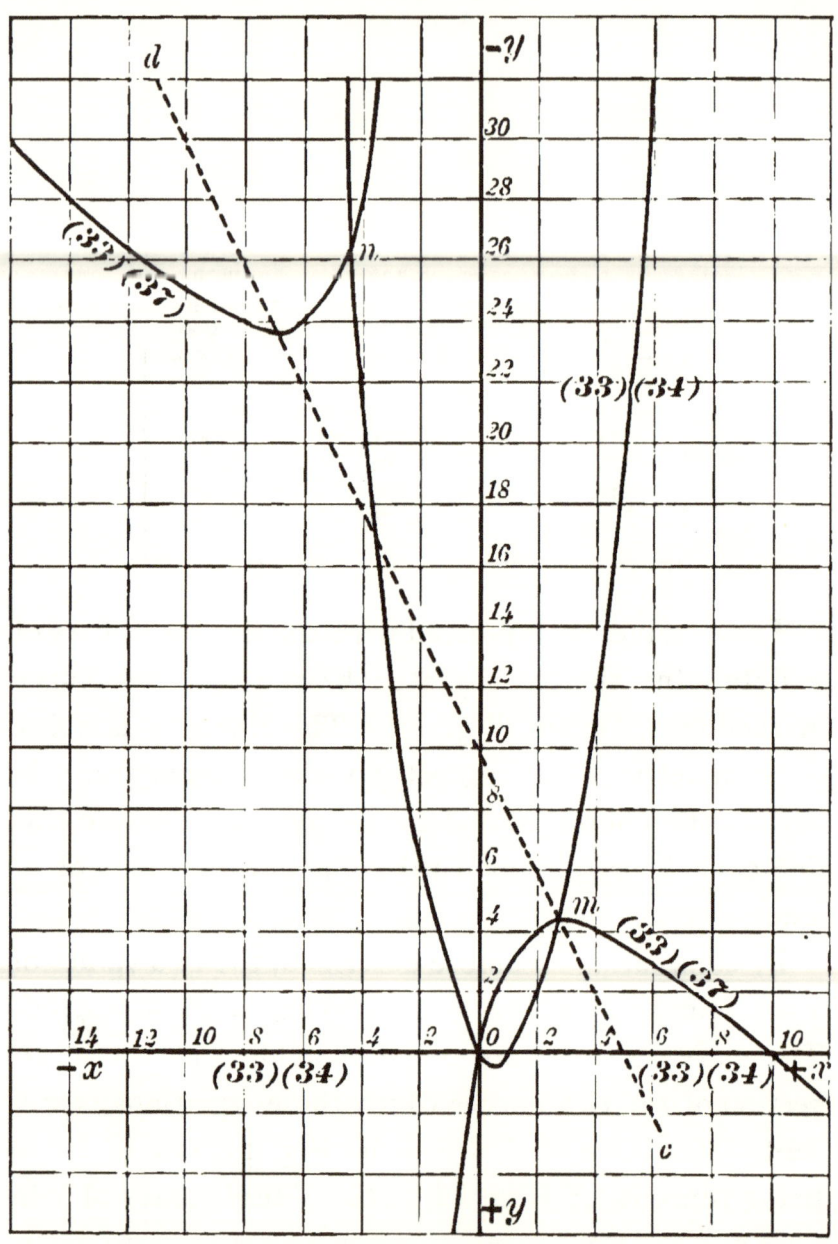

Fig. 17.

point of intersection of the two curves which is marked m in Fig. 17. The values $x = -4.4886$, $y = -26.1324$ locate the point marked n.

If the first set of values be substituted in either of the three equations (33), (34), or (37), the value of z is found to be $z = +2.9772$. The point common to the three surfaces is therefore above the reference plane x, y, a distance 2.9772. Placing the other set of values in either of the three equations we find $z = -11.6438$. This common point is therefore below the plane x, y by this distance. We observe also that both curves intersect at the origin, where $x = 0$ and $y = 0$. These values of x and y being placed in (33) and (34) will satisfy those equations for any value of z. The geometrical meaning of this is that the axis z lies in both of these surfaces. It is a line of intersection of these surfaces. Putting $x = 0$ and $y = 0$ in (37), we find that z must equal 10, and cannot have any other value when these conditions are imposed. Another point common to the three surfaces is therefore $y = 0$, $x = 0$, $z = 10$.

We may get additional evidence on this point by making $z = +10$ in the equations of the three surfaces. They then become

$$10x + 2y = 0, \quad \dots \quad (33)'$$
$$40x - 30y + 6xy = 0, \quad \dots \quad (34)'$$
$$x - y = 0 \quad \dots \quad (37)'$$

It will be seen that if $x = 0$ in any one of these equations, y is also zero. This indicates that on the horizontal plane where $z = 10$ the sections of the three surfaces cut through the axis z, x and y being then zero. If z be made any other value than 10, this will not then hold for equation (37), but it will hold for the other two equations.

By giving x various values from $+5$ to -5 in the three equations last written, the corresponding values of y may be computed, and the curves or lines represented by those equations may be drawn. This is done in Fig. 18.

The paper represents the plane $z = +10$. The axis z is a line at right angles to the paper, at the intersection of the axes x and y.

The three points common to the surfaces which we have found are

$x = 0,$ $\quad\quad y = 0,$ $\quad\quad z = +10,$
$x = +2.8219,$ $\quad y = -4.2009,$ $\quad z = +2.9772,$
$x = -4.4886,$ $\quad y = -26.1324,$ $\quad z = -11.6438.$

It is evident that by making $z = +2.9772$ or -11.6438 in the three fundamental equations, we may compute and draw the curves representing the sections of the surfaces on either of the horizontal planes thus determined, as has just been done for the plane $z = 10$.

Furthermore, by making $y = -4.2009$ or -26.1324 in the fundamental equations, similar sections on the vertical planes thus determined may be

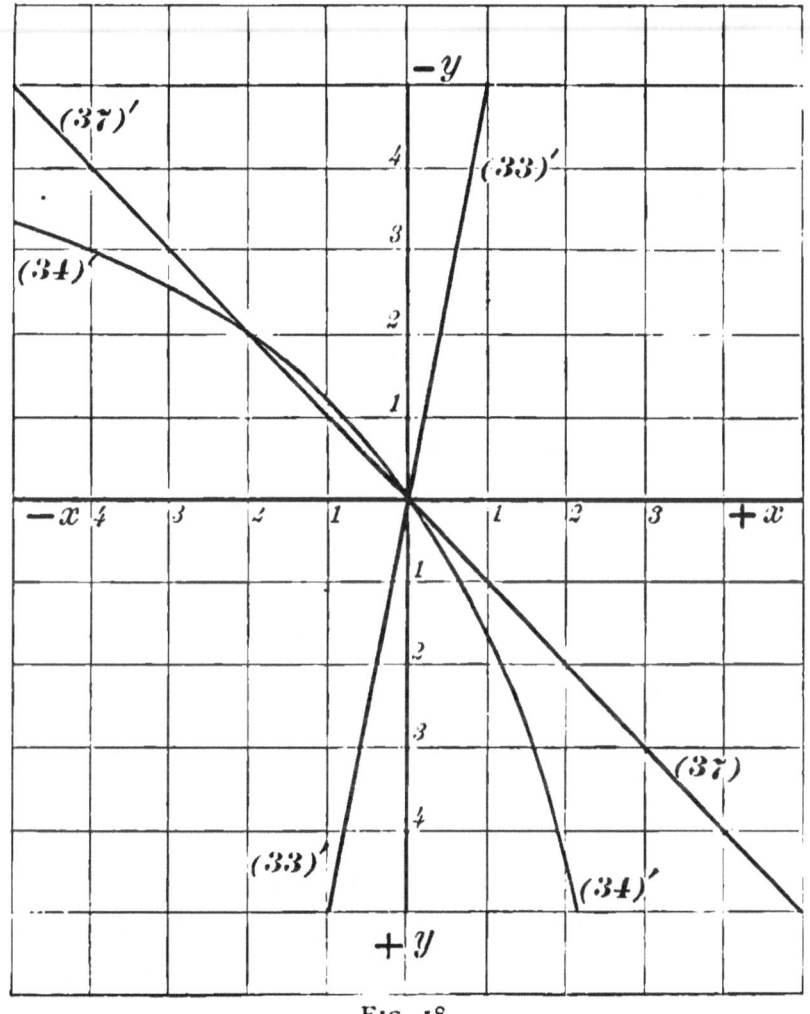

Fig. 18.

drawn. These planes are at right angles to the y-axis. The sections of the surface will thus be found to intersect at the points determined by the values x and z

above given. By giving values to x of $+2.8219$ and -4.4886 the sections at right angles to the x-axis through the points common to the three surfaces may be examined.

The pupil should not be disturbed that he is unable to form a complete mental picture of these surfaces and of their intersections. The methods which have been suggested will enable him to do so if he cares to persist and do the necessary computing for the eight trihedral angles determined by the reference planes. It should, however, be remembered that such discussions are continued in the college course, and that there will always remain many things yet to be learned.

17. It is easy, by means of four equations having four variable unknown quantities, to find values for the quantities which will satisfy all of the equations. The geometrical meaning of such equations has taxed the powers of the wise men for many years. Much has been written about the properties of space of four dimensions. But it does not appear that any one has been able to form any satisfactory physical or geometrical conception of what space of four dimensions might be. Here algebra and geometry appear at present to part company. And if that kind of space should come to be understood, we should still be in the dark concerning space of five or six or ten dimensions.

www.ingramcontent.com/pod-product-compliance
Lightning Source LLC
Chambersburg PA
CBHW020243090426
42735CB00010B/1816